EVENTS THAT CHANGED YOUR WORLD

ALEXANDER FLEMING DISCOVERS PENICILLIN

by Marcia Amidon Lusted

PEBBLE

a capstone imprint

Published by Pebble, an imprint of Capstone
1710 Roe Crest Drive, North Mankato, Minnesota 56003
capstonepub.com

Library of Congress Cataloging-in-Publication Data is available on the Library of Congress website.

ISBN: 9780756581138 (hardcover)
ISBN: 9780756581343 (paperback)
ISBN: 9780756581190 (ebook PDF)

Summary: How can medicine that tastes like bubble gum make you feel better so quickly? It has penicillin in it—an antibiotic that helps treat bacterial infections. But we wouldn't have the drug without the work of Alexander Fleming. Learn about the lifesaving impact of penicillin on the practice of medicine.

Editorial Credits
Editor: Ericka Smith; Designer: Terri Poburka; Media Researcher: Svetlana Zhurkin; Production Specialist: Katy LaVigne

Image Credits
Alamy: Chronicle, 13, Photo 12, 23; Associated Press: 21; Getty Images: Bettmann, 9, Hulton Archive/Baron, 5, 14; Library of Congress: cover (bottom), 16; Newscom: akg-images, 22; Shutterstock: Everett Collection, 10, 11, i viewfinder, 27, Jirawan muangnak, 15, Kallayanee Naloka, 17, Peakstock, 26, Prostock-studio, cover (top); SuperStock: Glasshouse/Circa Images, 7, Science and Society/SSPL/DHA/NMPFT, 19, Science and Society/SSPL/Science Museum, 25

Printed and bound in the USA. PO 5853

TABLE OF CONTENTS

Words in **bold** are in the glossary.

A Happy Accident

In the 1920s, scientist Alexander Fleming was studying **infections** that made people sick. He wondered what caused them. And he wanted to know how to stop them.

In the summer of 1928, Fleming stopped working in his **laboratory** to go on vacation. But he forgot to put away a dish of **bacteria**. Fleming's accident would lead to a big discovery—the **antibiotic** penicillin. And this discovery would change medicine.

Alexander Fleming

It Starts with Bacteria

People have always gotten sick. Sometimes bacteria make them sick. Bacteria live in our mouths and throats. They live on our skin too. Most are not harmful. But some can cause infections.

Before the discovery of penicillin, doctors could not always help people with infections get better. They tried many things. They used metals like mercury and silver. They also used **serums** made from animals' bodies. But none worked well.

A doctor examining a girl's throat in 1917

Doctors could not find the right medicine to fight bacteria. If people were not healthy, they could not always fight off the bacteria by themselves. So they did not always get better. Children had the hardest time battling infections. Their bodies were not as strong as adults' bodies.

Did You Know?

The body has an immune system that is made up of cells, tissues, and organs. They work together to protect the body. Children's immune systems are not as strong because they are not yet fully developed.

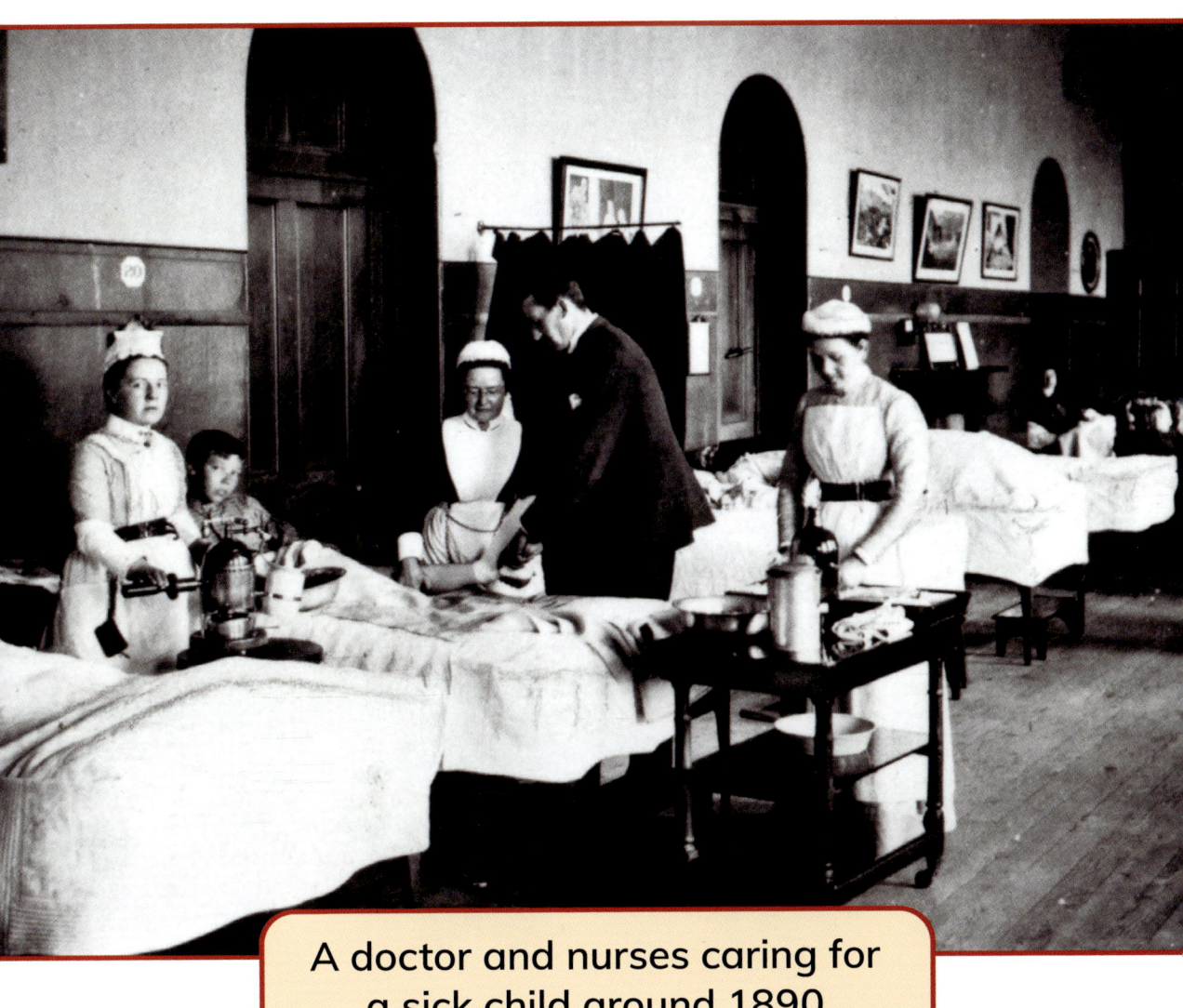

A doctor and nurses caring for a sick child around 1890

Looking for a Cure

During World War I (1914–1918), there was a great need to cure infections on the battlefield. Soldiers' wounds often became infected. Many soldiers died from the infections. Even getting a small splinter could cause a deadly infection.

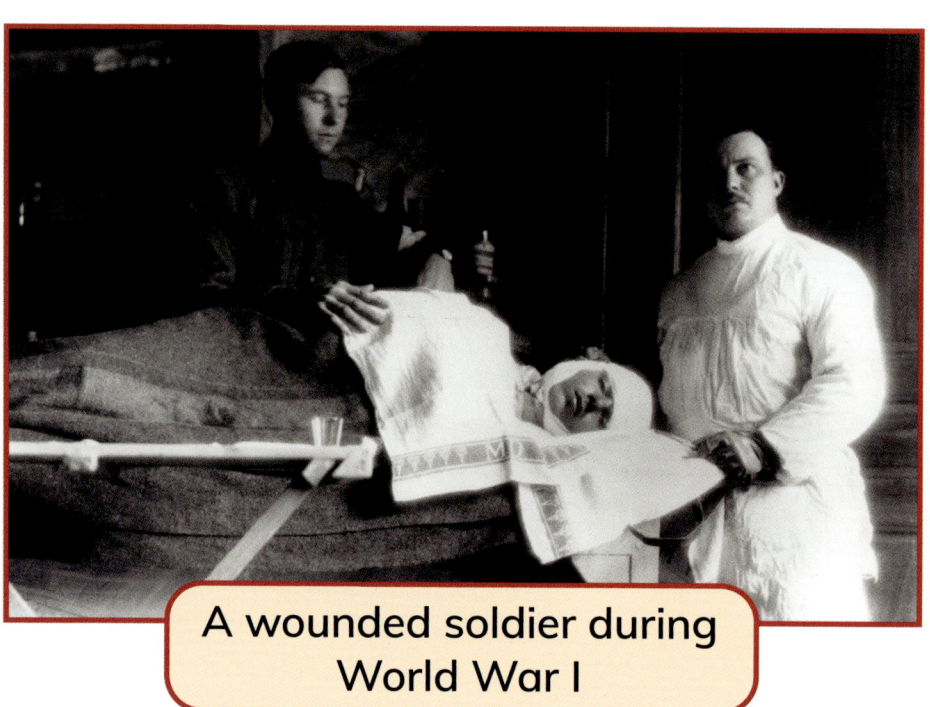

A wounded soldier during World War I

Soldiers also got sick. Many caught the flu. And some got an infection in their lungs called **pneumonia**.

There just were not enough medicines to help the soldiers.

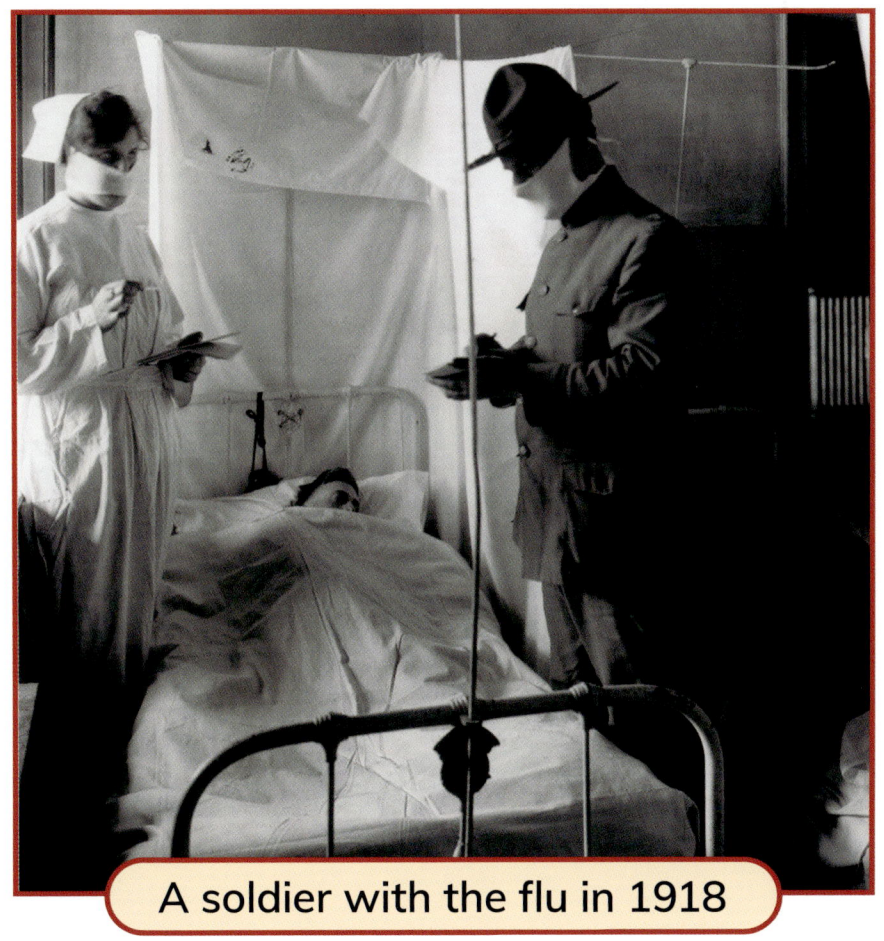

A soldier with the flu in 1918

Alexander Fleming was a young scientist from Scotland. He served during the war. He worked in hospitals near the battlefields. He saw many soldiers die from infection. He wanted to find a medicine that could keep soldiers from getting infections.

After the war ended, Fleming went back to the laboratory. He started working at St. Mary's Hospital in London, England.

Fleming in his laboratory in 1951

Fleming studied bacteria that caused terrible infections. People with weak bodies often got very sick from them.

In the summer of 1928, Fleming went on vacation. He accidentally left a sample of the bacteria sitting on his workbench when he left.

Fleming came back two weeks later. While cleaning his workbench, he found the sample of bacteria in a **petri dish**. But something had changed. There was **mold** growing in the dish. It was called *Penicillium*.

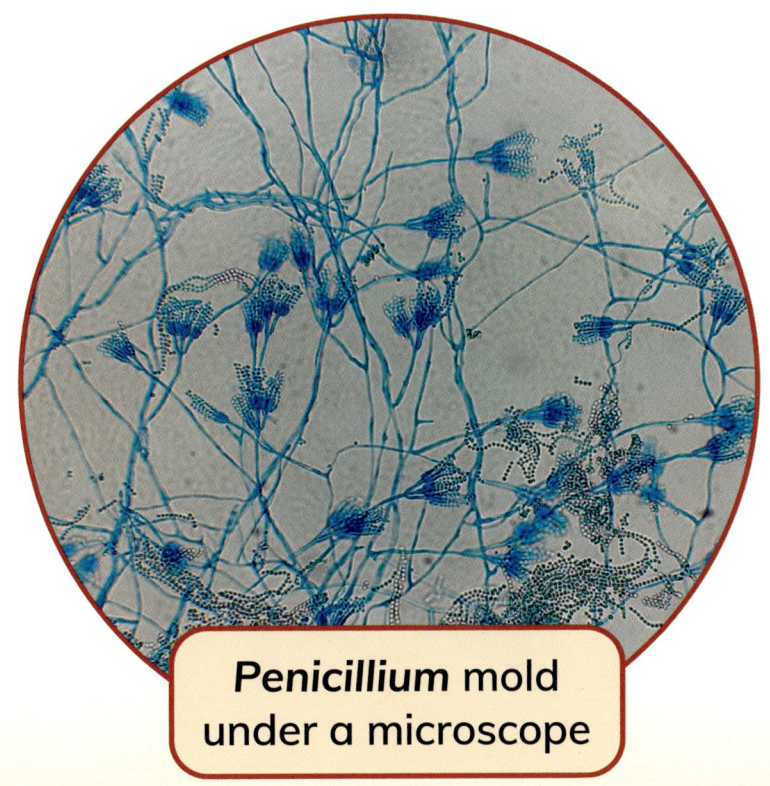

Penicillium mold under a microscope

Fleming looked at it closely. He saw that the area around the mold was clear. There were no bacteria there. Somehow a piece of mold had floated into the laboratory. It had landed in the dish and killed the bacteria that it touched.

Fleming in his laboratory in 1943

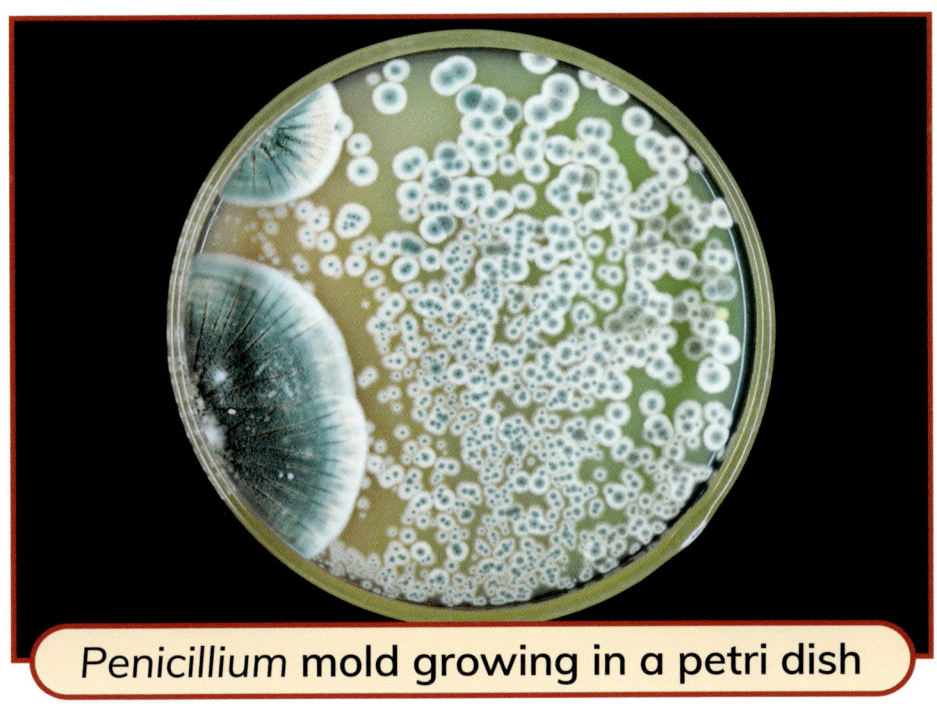

Penicillium **mold growing in a petri dish**

The mold had created a substance that killed the bacteria! Fleming named this substance *penicillin*. He discovered that it killed other kinds of bacteria too.

Did You Know?

Penicillin is an antibiotic. Antibiotics can kill bacteria. Some work on many kinds of bacteria. Some work on just a few.

But Fleming could not find a way to turn the penicillin into medicine. He needed to make a liquid from the substance. But it was hard to do, and he could not produce enough to test on people. He decided to work on other projects instead.

About 10 years later, scientists Howard Florey and Ernst Chain found Fleming's research. They did their own **experiments**. They figured out how to make pure penicillin from the mold. They also made sure that penicillin was safe for people. Now penicillin could be used as a medicine.

Working making liquid penicillin in a laboratory in 1943

Did You Know?

In 1945 Alexander Fleming, Howard Florey, and Ernst Chain received the Nobel Prize for discovering penicillin and making it a medicine. The prize honors people around the world who have done important work.

The Wonder Drug

One of the first people doctors treated with penicillin was a woman named Anne Miller. In March 1942, she was in the hospital. She had a very bad infection. She was dying. Doctors had tried many kinds of medicines to cure her. But none worked.

A doctor got a small sample of penicillin. They gave it to her as a shot. Her fever started to go away overnight. Soon she was feeling much better.

Miller met Fleming (right) in 1945.

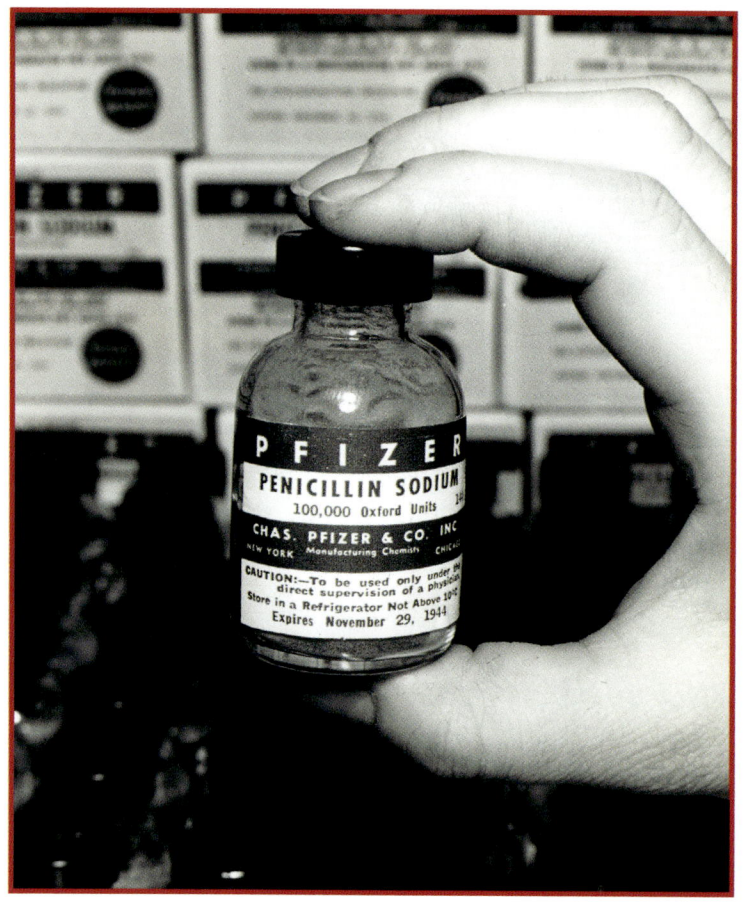

Doctors everywhere began using penicillin to cure infections. People started calling penicillin the "wonder drug." Just a little bit of it could cure serious infections.

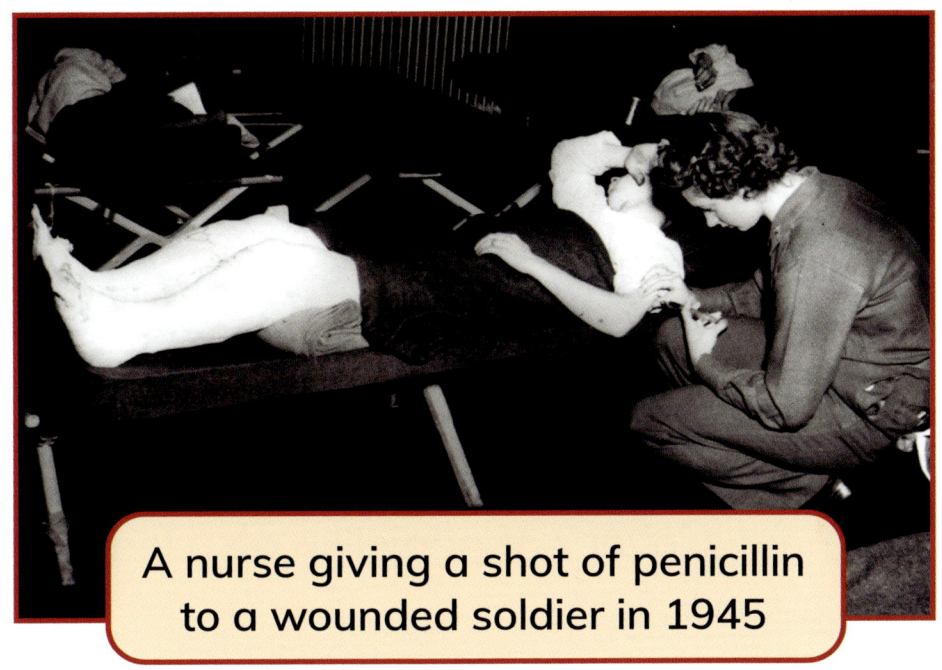

A nurse giving a shot of penicillin to a wounded soldier in 1945

During World War II (1939–1945), penicillin saved many lives. During World War I, only 4 percent of soldiers survived illnesses and infections. About 50 percent of soldiers in World War II survived them. Penicillin kept wounded soldiers from getting infections. And it helped them recover from illnesses like pneumonia.

After the war ended, access to penicillin increased. There was more of it available since soldiers no longer needed it. It saved many lives.

The discovery of penicillin also led to the creation of more antibiotics. Scientists discovered more substances that worked in similar ways. They used the same method used to create penicillin to create new medicines.

Workers making penicillin in the 1950s

Now, doctors have many kinds of antibiotics to give to patients. If one does not help you feel better, your doctor can give you another. Penicillin and other antibiotics help keep people healthy. Thanks to Alexander Fleming, your next earache or sore throat won't last long!

Did You Know?

There is a problem with antibiotics like penicillin. They can be used too often. Certain bacteria learn to resist the medicine. So doctors have to be careful and prescribe antibiotics only when they're really needed.

Timeline

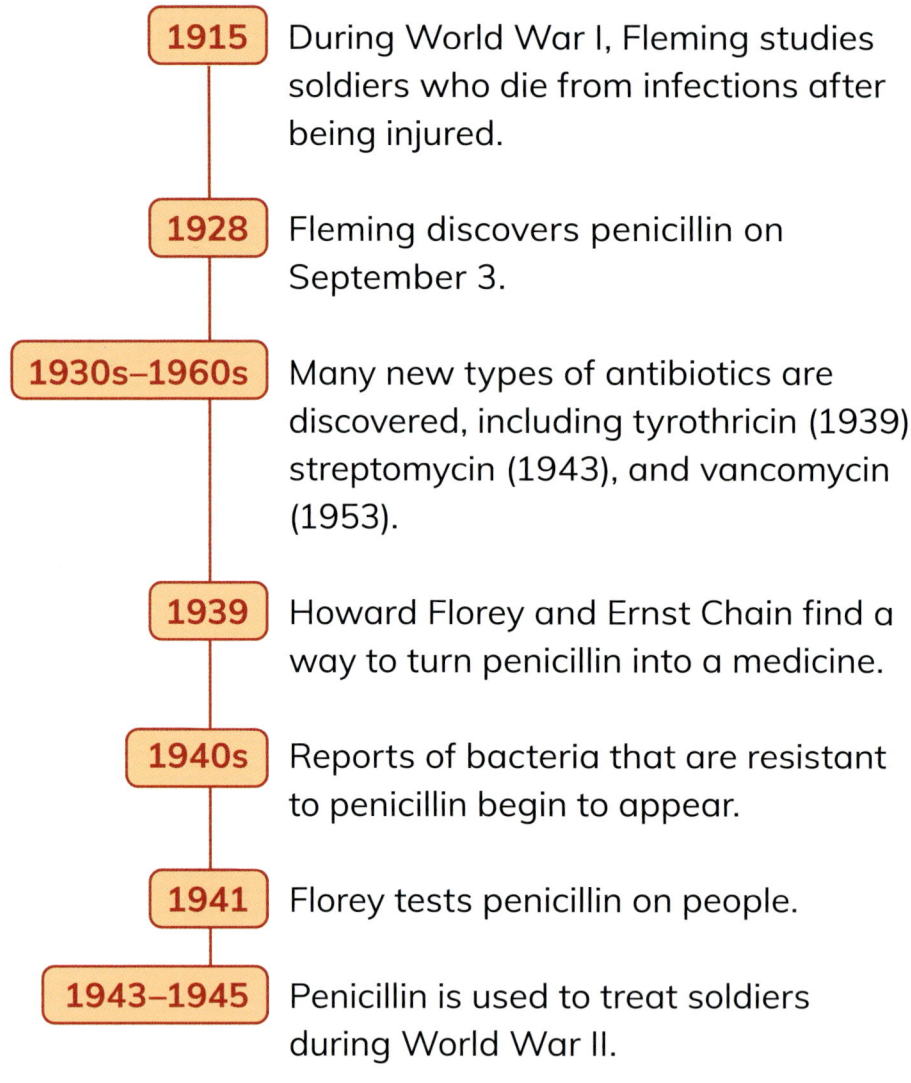

1915 During World War I, Fleming studies soldiers who die from infections after being injured.

1928 Fleming discovers penicillin on September 3.

1930s–1960s Many new types of antibiotics are discovered, including tyrothricin (1939), streptomycin (1943), and vancomycin (1953).

1939 Howard Florey and Ernst Chain find a way to turn penicillin into a medicine.

1940s Reports of bacteria that are resistant to penicillin begin to appear.

1941 Florey tests penicillin on people.

1943–1945 Penicillin is used to treat soldiers during World War II.

1945	Penicillin becomes available to everyone in the United States.
1945	Fleming, Florey, and Chain win the Nobel Prize in medicine for discovering penicillin and creating a medicine.
1955	Fleming dies.
2015	President Obama introduces a plan to address bacteria that are resistant to antibiotics.

Glossary

antibiotic (an-ti-bye-OT-ik)—a drug that kills bacteria

bacteria (bak-TEER-ee-uh)—very small living things; some bacteria cause disease

experiment (ik-SPEER-uh-muhnt)—a scientific test to find out how something works

infection (in-FEK-shun)—an illness caused by germs such as bacteria or viruses

laboratory (LAB-ruh-tor-ee)—a place where scientists do experiments and tests

mold (MOHLD)—a fuzzy substance that sometimes grows on old food; penicillin is made from a mold

petri dish (PEE-tree DISH)—a small, shallow dish used to grow things like bacteria

pneumonia (noo-MOH-nyuh)—a disease that causes the lungs to become inflamed and filled with fluid

serum (SIHR-uhm)—a liquid made from a fluid from an animal

Read More

Bard, Jonathan and Mariel Bard. *Oops! It's Penicillin!* New York: Gareth Stevens, 2020.

Capps, Heather Murphy. *The Amazing History of Medicine.* North Mankato, MN: Capstone, 2023.

Terrell, Brandon. *Antibiotics: A Graphic History.* Minneapolis: Graphic Universe, 2022.

Internet Sites

Britannica Kids: Penicillin
kids.britannica.com/kids/article/penicillin/390831

Fun Kids: Top 10 Facts About Antibiotics!
funkidslive.com/learn/top-10-facts/top-10-facts-about-antibiotics

Kids Discover: Penicillin: Who Found This Functional Fungus
kidsdiscover.com/quick-reads/penicillin-found-functional-fungus

Index

About the Author

Marcia Amidon Lusted has written over 200 books and 600 articles for young readers of all ages. She also writes and edits for adults and works in sustainable development. Visit www.adventuresinnonfiction.com for more about her books.